ISBN 978-1-333-43183-9
PIBN 10503799

This book is a reproduction of an important historical work. Forgotten Books uses
state-of-the-art technology to digitally reconstruct the work, preserving the original format
whilst repairing imperfections present in the aged copy. In rare cases, an imperfection in
the original, such as a blemish or missing page, may be replicated in our edition. We do,
however, repair the vast majority of imperfections successfully; any imperfections that
remain are intentionally left to preserve the state of such historical works.

1 MONTH OF
FREE
READING

at

www.ForgottenBooks.com

By purchasing this book you are eligible for one month membership to ForgottenBooks.com, giving you unlimited access to our entire collection of over 1,000,000 titles via our web site and mobile apps.

To claim your free month visit:

www.forgottenbooks.com/free503799

Correction of Video Camera Response
Using Digital Techniques

by

C. Marc Bastuscheck

Technical Report No. 285
Robotics Report No. 105
March, 1987

New York University
Dept. of Computer Science
Courant Institute of Mathematical Sciences
251 Mercer Street
New York, New York 10012

Work on this paper has been supported by Office of Naval Research Grant N00014-82-K-0381, National Science Foundation CER Grant DCR-83-20085, and by grants from the Digital Equipment Corporation and the IBM Corporation.

Correction of Video Camera Response using Digital Techniques

C. Marc Bastuscheck

Robotics Activity
Courant Institute
New York University

ABSTRACT

Video camera intensity non-linearities can be corrected without the use of light measuring equipment or calibrated filters. The camera to be corrected is used to indicate that two independent light sources illuminate a test card equally. Digitized images made using these sources separately and together are analyzed to determine parameters for use in generating correction tables. An example is presented in which a ±2% non-linearity observed in a GE2507 solid state camera was reduced to less than ±0.2% using a least squares fit to a third order polynomial. The technique can be used with camera output having any function (e.g. linear, logarithmic), and should be useful whenever strict adherence to a nominal output function is desired. Simple procedures for determining that a camera is non-linear and for correction of vignetting are also described.

1. Introduction

The output of a video camera only approximately follows its presumed functional form, e.g. linear or logarithmic in light intensity. In some applications deviations may be excessive, and correction functions must be determined and used. Perhaps the most exhaustively calibrated cameras are those which were flown to the moon and to Mars: sensitivity curves for each pixel were determined as functions of wavelength and temperature [1] using special light sources and measuring equipment. While modern cameras, and particularly solid state cameras, can be considered linear for many applications, even these show non-linearities if measured sufficiently accurately. It is widely recognized that the output of a camera can be corrected [2], but if measurements of light intensity are required the collection of data can be tedious. In this paper a method is described by which all data required to generate correction functions can be gathered using only the camera under test. Although deviations of the camera output from the desired form will be called "non-linearities", the techniques described here also apply to cameras with non-linear functions.

The camera to be tested is presumed to be used with a frame grabber, i.e. the output video signal can be sampled at video rates and stored digitally. The practical advantage of this is that video signals can be measured to better than one part per thousand by averaging many successive frames. The camera signal is represented as an array of numbers, and the camera itself can be thought of as an array of "light sensors", each of which has an output which is a function of the light incident upon

Work on this paper has been supported by Office of Naval Research Grant N00014-82-K-0381, National Science Foundation CER Grant DCR-83-20085, and by grants from the Digital Equipment Corporation and the

it. Note that these "light sensors" do not necessarily correspond to the sensing elements in a solid state camera, but rather to the picture elements (pixels) of the digitized image which depend on the digital sampling rate. Since the procedure described in this paper corrects the digitized output of the camera, non-linearities in the digitizing unit are also corrected. For purposes of discussion it will be assumed that the camera is non-linear.

The intensity I_{ij} of light incident on the "sensor" in the ith row and jth column is related to the corresponding output value O_{ij} by

$$O_{ij} = f_{ij}(I_{ij}), \tag{1}$$

where the function f may vary with position on the camera retina.

The response at each pixel must be independent of the response at other pixels. This is often not the case. For instance, if a camera has an automatic gain control the value at any pixel depends on the intensity at *all* other pixels. There may also be "blooming" of bright areas into adjacent dark areas or reflections or flare from poorly constructed or dirty lenses. However, cameras exist which approach ideal behavior, and in the following it is assumed that the output from each pixel is a function only of the light incident upon the corresponding area of the sensor.

In a useful camera f will be smooth and monotonic, and if both I and O are known (i.e. measured) many techniques can be used to determine f and its inverse, the correction function. Sawchuk [3] has considered a number of implementations by which correction tables may be generated, stored, and used to correct images in real time. In addition, if I is taken to be the light which is *expected* to be incident on a pixel (i.e. it is assumed that a uniformly illuminated surface will form an image of uniform intensity) the same machinery can be used to correct for the usual vignetting effect of lenses. A real-time correction scheme for this has been described by Onoe et al [4].

If f is the same over the entire sensor the requirements for data and computation are dramatically decreased, since only one correction function is needed. Theoretical and practical examples are given below for the case in which f is constant over all of the image. However, if f were different for each pixel or for regions of pixels the theory and light measuring technique would be changed only by collecting data and solving for a correction function for each region.

Implicit in eq. 1 is the assumption that both I and O are known for every i,j. While the camera output is easily measured, determining I at each region of the sensor is difficult. If a light measuring device is used it must be calibrated and it must be carefully used. Care must also be taken when changing light intensity that the color temperature is not changed as well, since the spectral response of a lightmeter is not likely to match that of the camera.

Knowledge of absolute intensity is in fact not needed to linearize a camera response. The intensity of a constant light source could be modified in a known manner to provide many different but related intensities. For example, a point source could be placed at different distances from the camera retina (exposed with no lens), but this requires a large dark room for adequate dynamic range, and may

---

IBM Corporation.

entail some difficulty in defining a "point" source and measuring its location at close separations. Neutral density filters or mirrors [5] having special properties could be used, but these must be calibrated, properly maintained (no dust or scratches), and carefully used.

The method introduced here is to arrange two light sources so they appear to the camera to illuminate a surface identically when used separately, then record the observed intensities for the sources used separately and together. Although the absolute intensities cannot be determined, it is known that they have the ratio 1:2, and these observations, made for many intensities, are sufficient to discover a function which will linearize the output. There is one proportionality constant which is not determined by this method, and it can be chosen so the (linearized) output function spans the available digits. Precautions must of course be taken to control ambient light, and also to ensure that the two light sources do not interact with each other. For example, if two lamps share a power line, turning one of them off will usually cause the other to brighten.

2. Theory

It is assumed that measurements can be made, for example using the 'two light' protocol described above, for which the true intensities are known to be in a ratio of 1:2. A functional form relating the observed camera output O to the intensity on the retina I is assumed. The exact form chosen depends on the expected camera function and on the form of the observed deviation; some trial and error may be necessary to find a suitable parametrization. It is convenient to write this as $I = C(O)$ where C would be the inverse of f above. (In the general case I, C, and O would have subscripts i,j.) Equations are then written for two intensities differing by a factor of 2, and the absolute intensity is eliminated between these equations to leave an expression containing only the parameters of the function, the observations, and the factor of 2. This expression can then be used to find the values of the parameters. Two examples follow, one in which C has an exponential form and one in which C is a general polynomial expansion.

Dinstein et al [5] corrected the output of their video tube camera using an equation of the form

$$I = b(O - d)^{\alpha} \qquad (2)$$

where b is a constant, d is the dark image, I is the true intensity, and O is the observed intensity. It is assumed that the constant d can be determined, for example, by making an image with the lens cap on. Suppose that intensity I gives O_1 and $2I$ gives O_2. That is,

$$I = b(O_1 - d)^{\alpha} \qquad 2I = b(O_2 - d)^{\alpha}. \qquad (3)$$

Taking the ratio yields

$$2 = (\frac{O_2 - d}{O_1 - d})^{\alpha} \qquad (4)$$

and taking the log of both sides gives

$$\alpha = \frac{\log 2}{\log(O_2 - d) - \log(O_1 - d)}. \qquad (5)$$

From this equation and a small number of measurements α can be determined. Of course, if α is not constant for all intensities a different form for the correction function must be assumed. The constant b can be arbitrarily chosen, e.g. to make the corrected intensities span the range of digits available in the image processor.

A more general approach is to determine parameters for a correction function $C(O)$ expressed as a polynomial in the observations O:

$$C(O) = a + bO + cO^2 + dO^3 + \cdots = b(\alpha + O + \gamma O^2 + \delta O^3 + \cdots \qquad (6)$$

where a,b,c,d,α,γ and δ are constants and would have subscripts i,j if C varied across the image. For a particular pair of observations O_1 (using a single light source) and O_2 (using both sources) it is known by assumption and experimental arrangement that

$$\frac{C(O_2)}{C(O_1)} = 2.00, \qquad (7)$$

or

$$\alpha + O_2 + \gamma O_2^2 + \delta O_2^3 = 2(\alpha + O_1 + \gamma O_1^2 + \delta O_1^3) \qquad (8)$$

or

$$\alpha + \Theta_1 + \Theta_2 \gamma + \Theta_3 \delta = 0 \qquad (9)$$

where

$$\Theta_1 = 2O_1 - O_2 \qquad \Theta_2 = 2O_1^2 - O_2^2 \qquad \Theta_3 = 2O_1^3 - O_2^3. \qquad (10)$$

Note that the Θ's are combinations of observed quantities.

This formulation lends itself to least squares determination of the constants. If the expression on the left hand side of eq. 9 is the deviation q from zero of any particular experimental set of observations, then the square of this expression summed over all observations is a measure of the error. The constants α, γ, and δ must be chosen to minimize this error, and equations to this end are generated by the constraints that partial derivatives of $\sum q^2$ with respect to α, γ, and δ must equal zero. This method is easily extended to higher order approximations in the expansion for $C(O)$, and additional equations are generated for each higher power at the cost of a larger set of linear equations to be solved. These equations are written easily in terms of Θ's, which are combinations of measured quantities. For the third power expansion the equations are:

$$\alpha \sum 1 + \gamma \sum \Theta_2 + \delta \sum \Theta_3 = -\sum \Theta_1 \qquad (11)$$

$$\alpha \sum \Theta_2 + \gamma \sum \Theta_2^2 + \delta \sum \Theta_2 \Theta_3 = -\sum \Theta_1 \Theta_2$$

$$\alpha \sum \Theta_3 + \gamma \sum \Theta_2 \Theta_3 + \delta \sum \Theta_3^2 = -\sum \Theta_1 \Theta_3$$

where the sums are taken over all observations. The constant b of eq. 6 cannot be determined, but (as in the exponential case) can be chosen arbitrarily for convenience.

- 4 -

3. Example: linearization of a G.E. 2507 solid state camera

Inconsistencies in combinations of images generated by a General Electric model 2507 solid state surveillance camera suggested that the camera output was non-linear. Preliminary measurements were made using two light sources, and it was quickly determined that non-linearities existed and that a simple exponential form (eq. 2) was inappropriate. Preliminary experiments also indicated that a single correction function could be used for all areas of the image.

The existence of non-linearities can be demonstrated easily using two light sources. The camera is focused on a surface having a sharp change of reflectivity, for instance a gray card mounted on a white board. Two light sources of different intensity are arranged, such as a projector with neutral density filters or a pair of lamps. To reduce the effects of specular reflectivity the surfaces should have low gloss and the light sources should be at the same angle to the surface. Note that if colored paper is used for the change in reflectivity the light sources must have the same color balance. An image A is made using ambient light; an image I_1 is made using the first source, and I_2 is made in the light of the second source. The ratio image

$$R = \frac{I_1 - A}{I_2 - A} \tag{12}$$

is made by carrying out subtractions and division for corresponding pixels in the images. A change in the ratio image at the change of reflectivity indicates a non-linear response. The experiment should be carried out for several different absolute intensities (e.g. by changing the camera diaphragm setting) because absence of a change at one intensity does not guarantee linearity at all intensities.

A more elaborate experiment was used to determine the linearizing correction function. The camera with a good quality 50mm lens was pointed toward a matt white Formica screen placed perpendicular to the camera axis at a distance of about 1 meter. The screen was illuminated using graded neutral density filters to modify the light from two Kodak Ektagraphic slide projectors which were placed one above the other 1.3m to the right of the camera. Nearly identical patterns of illumination were achieved by adjusting the projector positions, directions and zoom lenses, providing a method of conveniently generating an intensity and its double at all locations in the image. The carousel style slide projectors were under computer control; when advanced a filter was projected, when subsequently reversed the beam was blocked by the mechanism so the lamps burned all the time.

The basic projector installation was not changed during the measurements. To cover the entire dynamic range of the camera the intensity at the retina was changed by opening or closing the diaphragm. To provide observations at a large number of different intensities neutral density filters whose transmittance varied smoothly by about a factor of two across the field of view were used. To reduce the effects of dust on the filters the projectors were focused at a point halfway to the screen. Lights in the room were turned off, and ambient light observed by the camera varied from less than 1 unit (of 2000 full scale) to about 5 units at the most open f/ stop. The ambient light was subtracted (with the dark image) from all images made, and had negligible effect on the linearization calculation. It is important that ambient light be as small as possible.

At a given diaphragm setting the ambient light was digitized, then first intensity, second intensity, and combined intensities. Each digitization was the average of 8 images to reduce noise. The average of the magnitude of the difference between consecutive (unaveraged) images was about 0.8% of full scale at moderate intensities. The ambient light image was subtracted from each of the other images, each image was condensed by averaging 16 pixel square blocks. The resulting set of three 32x32 pixel images was written to a VAX 750 for further work. All early processing was done using a VICOM image processing system which executed commands passed to it by a host VAX 750. Arithmetic operations were performed using 12 bits, including the sign bit. At each f/ stop (except the very darkest end) this image collection was repeated four times; in all 28 sets of images were collected. This large amount of data reduced the effects of random noise in the camera and temporal variation in the light sources.

Data was extracted from these reduced images by finding, in a set of images, a pair of corresponding 'single intensity' pixels within some ϵ of each other; then these values and the value of the corresponding pixel in the 'combined intensity' image were saved as a data set. Values of ϵ were chosen to produce roughly the same number of data sets from each exposure condition; ϵ varied from 5 (of 2000 full scale) at the high exposure end to 1 at the lowest. In all, 3500 data sets were extracted from the reduced images. These were checked to see if the images of one beam were perhaps consistently brighter than the other (since this could affect the correction function slightly under some circumstances). The average signed difference between the two single beam conditions was less than 0.2 times ϵ, showing that the patterns of illumination were very well matched.

A measure of the non-linearity of the camera-digitizing system is the observed "excess", i.e. the difference between the sum of the single beam digitizations and the value of the double beam digitization. In Fig. 1, this excess is plotted as a function of the observed value for both light sources. Each circle represents a single data set, and only a small number of the data sets are shown. If the system is linear no deviation from zero should be seen; in Fig. 1 variation between -20 and over 100 units (of 2000 full scale) is observed. The relation between Fig. 1 and the camera sensitivity- function is not simple, but the maximum amount of excess is roughly comparable to the maximum deviation from linearity. Thus one might expect to find the camera's deviation from linearity to be about 100/2000 or 5%.

The 3500 data sets were divided into five subsets, each of which spanned the range of observations, and the constants α, γ, and δ were determined using the least squares method given in the theory section. No significant differences were observed between correction functions generated using these different sets of constants: corrections made with different sets deviated from the average value by at most 2 units (2000 full scale) in the region of maximum deviation. Constants α, γ, and δ, determined by averaging the values obtained for each of the subsets, were used to generate a linearizing look-up table. Values used to generate the table were $\alpha = 10.036 \pm 0.19$, $\gamma = (-2.49 \pm 0.24) \cdot 10^{-5}$, and $\delta = (3.73 \pm .21) \cdot 10^{-8}$. Experimental uncertainties amount to approximately 0.1% of full scale at half of full intensity, where the error is largest. Note that in mid-range the cubic term of eq. 6 has a magnitude similar to that of the quadratic term.

The least squares procedure leaves a scale constant (b) undetermined, and this was set so that an observed value of 2000 returned a "true" value of 2000. Values in the look-up table can be used to estimate the non-linearity of the camera. For example, Fig. 2. shows the deviation from the true value plotted against the true value. In midrange deviations of nearly 80 units of 2000 full scale are observed; this could be expressed as a non-linearity of nearly 4% of full scale, or (using a different baseline), of ±2%. Although this is large enough to be troublesome in some applications, it is so small that if the camera output were plotted as a function of intensity the curve would be hard to distinguish from a straight line.

The linearization process can be evaluated by plotting the "excess" for corrected images. Fig. 3 shows plots of "excess" vs observed values in the manner of Fig. 1 for several different conditions. In Fig. 3a the data of Fig. 1 is plotted after it was corrected using a quadratic expression for $C(O)$ (i.e. eq. 6 without the cubic term). Fig. 3b shows this same data after correction using the cubic expansion described above: the "excess" varies by about ±10 from zero, suggesting a residual non-linearity of about 0.1%. Fig. 3c shows corrected data gathered three months after the linearization procedure was implemented. Again the "excess" differs from zero by only about 10 units, indicating that the improved linearity is stable in time.

4. Summary, Discussion, and Applications

A procedure by which the non-linear response of a video camera can be corrected without measuring light intensity or using calibrated equipment of any sort has been described. The camera under test is used to determine that light from two sources is of the same intensity, and the two sources are then used separately and together. The camera must be reliable to the extent that the output is reproducible and varies strictly monotonically with the intensity, and the output in one region of the image is entirely independent of all other regions of the image. While theoretical and practical examples were presented assuming that the output of the camera was to be made linear, very similar procedures can be used to make the output of a logarithmic camera exactly logarithmic, etc. The method of taking data would not change at all.

Several forms for correction functions were suggested above. However, the linearization procedure given here can be used with any function. A simple extension would be to replace eq. 6 with a polynomial of higher degree; in fact, the plot of the excess using the freshly computed correction table (Fig. 3b) suggests that higher powers might result in better linearization. However, if the camera non-linearities change in time (a possibility suggested by Fig. 3c) corrections using higher order terms may not be useful.

The limit to which a camera - digitizing system can be made linear is determined by the quality of measurements which can be made, the suitability of the form of the correction function, and the stability of the non-linearities, which may change with time, temperature, line voltage, etc. It appears unlikely that the camera tested above can be corrected to better than about 0.1 or 0.2% non-linearity, since at this level effects of random noise, temporal variation of light intensity, residual non-linearities beyond the compass of the cubic expansion, and variations in the camera behavior all seem to be equally important. However, this correction represents considerable improvement over the uncorrected + −2% non-linearity.

Slide projectors were used in the experiments described above because this seemed to be experimentally most convenient. Preliminary experiments were made using light bulbs instead of projectors, with a shielded "ballast" bulb to reduce the effect on the voltage of turning lamps on and off. It was possible to get ratios of 2/3, 1/3, 3/4, etc in place of the ratio 1:2 used here. However, control of ambient light was more difficult, and ordinary light bulbs were found to vary considerably in the amount of light they generate, the directionality of this light, and the amount of current drawn (which causes variation in the brightness of other lamps). Data provided by two matched sources is adequate for the method of generating a linearizing function given above.

Little time is required to generate a correction function. Software requirements are modest, consisting of a least squares solution of eqs. 11 or the equivalent and a few programs to read values out of images. The light source requirements can be met most easily using slide projectors, since these combine strong lights (allows the diaphragm to be less open, reducing the effects of ambient light) and excellent containment of that light (the ambient light should be small). The entire experiment can be carried out in several hours, allotted as one hour to set things up, and one to two hours to collect data. Computations require negligible time.

In the examples given above it was assumed that the one correction function would serve for the entire image area. In some cameras this is not the case, and the calculations which were made here once would have to be made for a number of different image regions. If a separate correction function is required for each pixel it may be more economical to compute images of the correction constants and sum the various terms later. Particularly if each coefficient does not vary greatly across the image this correction can be made using 12 bit arithmetic on an image processor which quickly adds or multiplies entire images. The current implementation of the "cubic" depth sensor computes depth from an observed ratio image using $Z = A + r*(B + r*(C + Dr))$ where A, B, C, and D are precomputed calibration images and r is the small deviation of the observed ratio from a calibration image of central ratio values R.

In principle the camera correction technique could be used to correct "vignetting" effects (i.e. more light is focused on the center of the retina than on the edges). However, it is generally much simpler to make an image of a white or gray card held at the position of the object and digitize an image of this - making sure that the lighting and lens settings, particularly the f/ stop, are the same as those used to digitize data images. The data images are then corrected by simply dividing them by the image of the card. This procedure is particularly useful for correcting images of printed material.

Correction tables or images can be used to correct images in real time. Although in some cases this may require specially designed hardware, much of this hardware either exists or will exist very soon in commercially available image processors which can be programmed to "pipe" an image through a number of processes in real time. For example, an image acquisition procedure might involve digitizing and storing a dark current image (lens cap on), and storing also an image made in the ambient light and corrected by subtracting the dark current image and then running the result through the correction look-up table; these images would be made as part of a set-up procedure. Subsequent data images might be made by digitizing in structured illumination, subtracting the dark current image, correcting the result (look-up table), and subtracting the image of the corrected ambient light. Each of these processes involves a single image location and is well suited to pipeline processing. There is no reason to wait for the last pixel of an image to be digitized before correcting the first pixel.

5. References

[1] W.B. Green and R.M. Ruiz, "Removal of photometric distortion from Mariner-9 television images," abstract in J. Opt. Soc. Am. **62** pp. 1351-1352 (1972).

[2] F.C. Billingsley, "Applications of digital image processing," App. Optics **9**, pp. 289-299 (1970).

[3] A.A. Sawchuk, "Real-time correction of intensity nonlinearities in imaging systems," IEEE Trans. Computers **C26** pp. 34-39 (1977).

[4] M. Onoe, M. Ishizuka and K. Tsuboi, "Real-time shading corrector for a television camera using a microprocessor," in *Real-time/Parallel Computing*, M. Onoe, K. Preston and A. Rosenfeld, eds., pp. 339-346, Plenum, New York, 1981.

[5] I. Dinstein, F. Merkle, T.D. Lam and K.Y. Wong, "Imaging system response linearization and shading correction," Opt. Eng. **23** pp. 788-793 (1983); also in Int. Conf. on Robotics, pp. 204-209, Atlanta (1984).

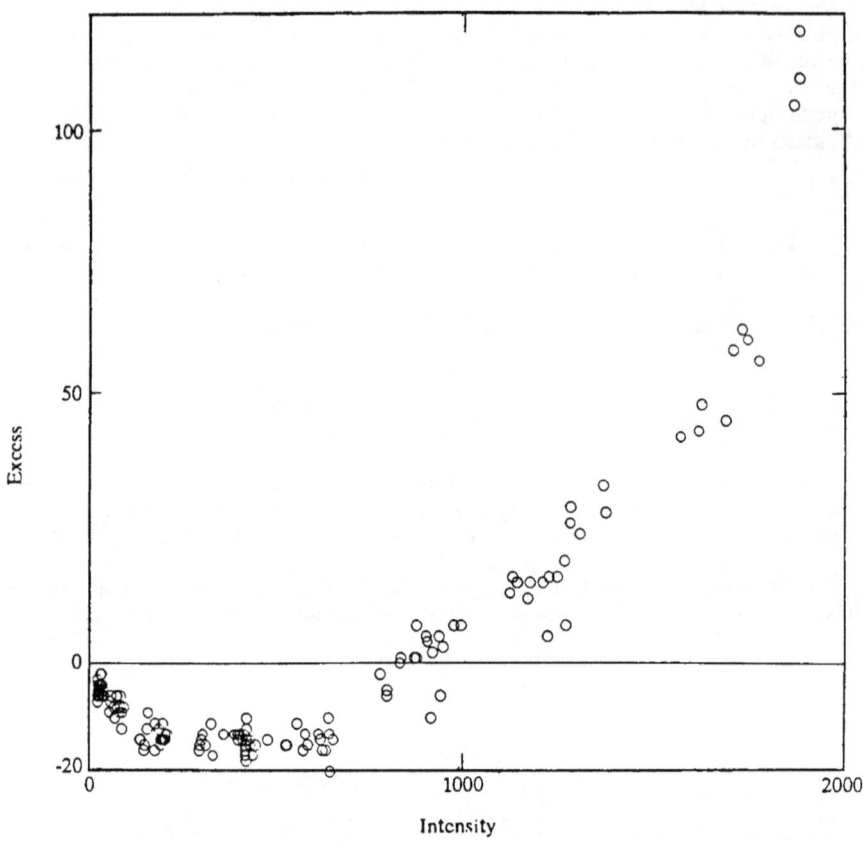

Fig. 1. Non-linearity of observed values as a function of observed value. "Excess" is defined as $2O_1 - O_2$, where O_1 is the value observed for intensity I and O_2 is the value observed for intensity $2I$. If the camera were linear the circles would fall on the horizontal axis.

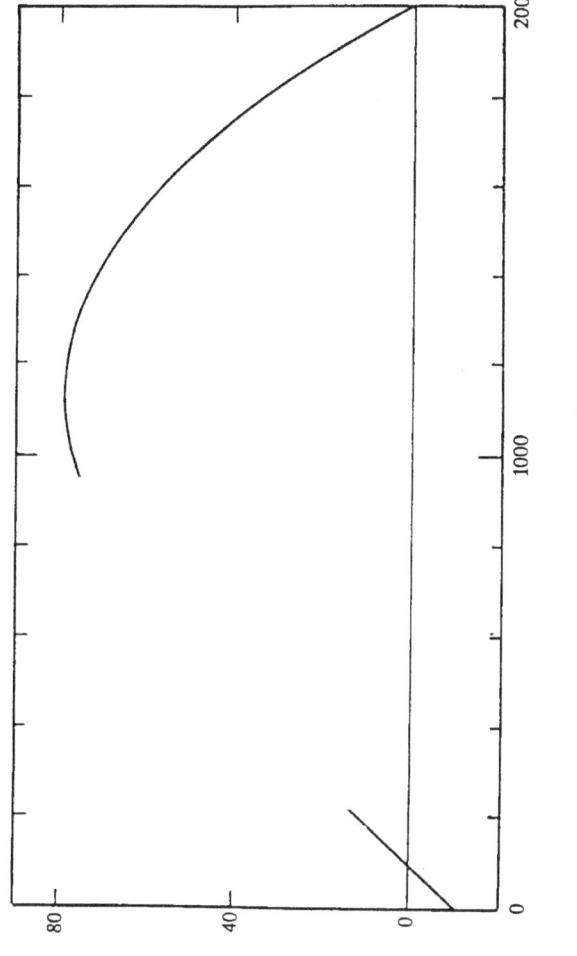

Intensity

Deviation from Linearity

Fig. 2. Deviation from linearity of the observed camera output values, plotted as a function of the corrected digitized intensity. The correction was generated from 30 sets of data similar to the data used to make Fig. 1 by determining coefficients for a third order polynomial function.

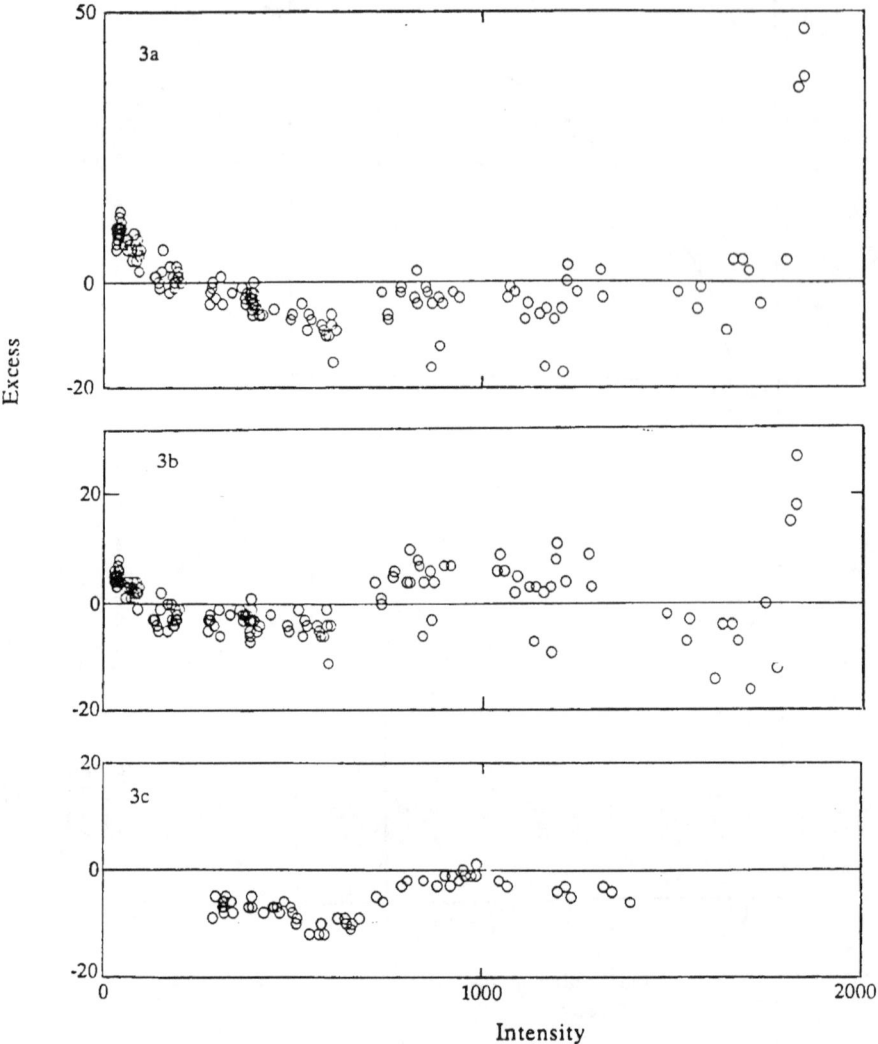

Fig. 3 Plots of the excess (as in Fig. 1) for several sets of corrected data. In Fig. 3a a quadratic correction has been applied to the data of Fig. 1. Fig. 3b shows the data of Fig. 1 corrected using the correction function plotted in Fig. 2. Fig. 3c shows data collected three months after the look-up table correction was implemented, demonstrating continued effectiveness.

CPSIA information can be obtained
at www.ICGtesting.com
Printed in the USA
BVHW041446190219
540527BV00069B/3947/P